# INTERNET YELLOW PAGES

# 99

# LINKS

## TO FIND A JOB

# WWW.99LINKS.ORG

# 2014

**ISBN-13:**
**978-1512192919**

**ISBN-10:**
**1512192910**

**99links to find a job – by BASTIAN LUCK**

# INDEX

# ABOUT 99LINKS

After some time looking for a job and taking down notes on all the websites I went thru, I decided to put together all this information on a book I titled _99Links to find a job_ with the hope of helping others on their search.

Immediately after 99LINKS to find a job was finished I decided to also use my notes and wrote down the second title – 99LINKS to fund your start-up. So this was the beginning of the book series 99LINKS.

99LINKS, a new book series containing a detailed list of the best websites for any given category.

We invite our readers and friends to create their own 99LINKS project and send it to us for evaluation. If your idea is good, then we can talk business.

For more details check our website:

www.99LINKS.ORG

If you have any suggestions or comments, please send us an email to: info@99links.org

# COMING SOON

We are almost finished with the next 99LINKS series book. We will dedicate it to list down the best 99 links to fund your start-up.

From Crowdsourcing to Venture Capitals, all the most popular and up to date links you are looking for.

ALSO we will launch 99 links to sell online and 99 links to find love.

## 99
### LINKS
**TO FUND YOUR STARTUP**

## 99
### LINKS
**TO SELL ONLINE**

## 99
### LINKS
**TO FIND LOVE**

# 99 LINKS ICONS

 These icon marks an information section. Usually about the background of a website or company.

 Keep an eye for this symbol, as it will indicate information about interfaces between networks.

 These icon will come together with a quick review on how things works with a company or website.

 We will flag with this symbol anything related to staff, people, location or amount of users.

 We selected some websites that are set up in the local language. So we will flag them with this symbol to let you know.

 We will sign up a page to recommend it.

# INTRODUCCION

In **99 Links TO FIND A JOB** we decided to divide our selection of web sites into 4 sections:

Starting with **Online CV Creators**, you will find a list of the most user friendly websites to generate you CV online or on a PDF format.

The second section it is all about **Online Professional Networks.** We all know LinkedIn and here you will find a great variety of offers to expand, centralize or specialize your profile online.

Our third section shows all inks to **Recruitment Agencies**. We shortlisted 25 of the most influential recruitment company not only in Europe but in the world.

Finally we review the **Online Job Offers**, from Monster.com to TotalJobs.com you will find the most popular websites with job offers online, not only in Europe but also in Middle East, America and Asia.

# ONLINE CV CREATORS

www.cvgram.com
www.about.me
www.cvmaker.com
www.visualcv.com
www.zerply.com
www.onlinecvgenerator.com
www.re.vu
www.talentous.com
www.resumeup.com
www.doyoubuzz.com
www.modelocurriculum.net
www.cuvitt.com
www.cvaudere.com
www.prezi.com
www.office.microsoft.com
www.vizualize.me
www.pdfcv.com
www.kinzaa.com
www.comoto.com
www.sliderocket.com
www.easy-cv.com
www.vizify.com
www.resumesimo.com
www.express-cv.com

# CVGRAM

Web address: **www.cvgram.me**

Headquarters: Argentina

Active since: 2012

A very relaxed and hipster styled platform. The Websites offers a simple way to create and edit your CV online.

It is possible to interface your Linkedin or Facebook profile information directly into the public profile form of CVGRAM, making it really easy to start shaping your infographic.

Although it is on its starting phase and lacks of some functionalities the website is cool and has a good customer service platform powered by *Get Satisfaction.*

**RECOMMENDED!**

# ABOUT.ME

Web address: **www.about.me**

Headquarters: San Francisco, USA.

Active since: 2009

Is a personal web hosting service co-founded by Ryan Freitas, Tony Conrad and Tim Young in October 2009.A great solution to promote your professional profile online and centralize all your contact channels into one platform.

With About.me you can redirect all your contacts from Facebook, twitter, wordpress, business cards or blogs to your online public profile.

Really easy to use and understand. Check out the blog section and read others review and profiles. It is quiet interesting how about.me could develop.

# CVMAKER

Web address: **www.cvmkr.com**

Headquarters: London, UK.

Active since: 2010

 With Cvmkr.com you can Create, maintain, publish, and share your CVs for free. A really good presentation and very easy to use. It is probably the most user-friendly CV generator available online.

 You can interface your Linkedin or facebook profile to create your login credentials. The website is available in English and 16 other languages.

 You will have to download google chrome to download the free templates, but that should not be an issue.

# VISUALCV

4

Web address: **www.visualcv.com**

Headquarters: Richmond, Canada.

Active since: 2008

Instead of finding the traditional structure to organize a CV, VisualCV offers you your very own personal webpage.

The most interesting aspect of this platform are the privacy settings, that aloud you to keep all your information privately or as public as you can, But there is no interface with Linkedin.

You can generate your professional profile, where you can add videos, images, presentations and graphics of your projects or on-going job.

# ZERPLY

Web address:           **www.zerply.com**

Headquarters:          Tallinn, Estonia

Active since:          2011

 Created by a group of designers in Tallinn Zerply generates in a simple but stylish way your CV. By May 2012 they raised over U$ 600k from a round including *500 Startups* and *Quotidan Ventures*.

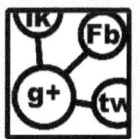 Using professional design tools that makes easy to create an visually attractive Resume. You can do everything you expect social wise, from follow ups to reviewing stats.

 The info graphic displays are smooth and clean, another added value you will find is the many ways to develop your distribution channels to promote your profile.

# CVGENERATOR

Web address:**www.onlinecvgenerator.com**

Headquarters:                USA

Active since:                2010

 If you need an smart and simple way to create your CV this is the easiest way to make CV online with their resume generator. Just answer few questions and the website will generate resume in pdf.

 An ideal solution for those who never did an online project or have little knowledge on Computers.

 It is so simple that even at the very end of the process the program will convert your CV into a PDF format and will save it directly onto your desktop.

# RE.VU

Web address: **www.re.vu**

Headquarters: California USA

Active since: 2011

 Follows the same concept we found in aboute.me linking the profiles of your other networks into one.

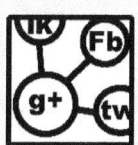 You can save a lot of time by importing all your data from Linkedin into Re.Vu. Only disadvantage is that you cannot export into PDF, but you can share your information with other contacts via Google+, Twitter, Facebook or Linked-in.

 So if your are looking for a clever way to centralize all your social and professional networks, this is your answer. Check out the wordpress comment feature.

# TALENTOUS

Web address:  **www.talentous.com**

Headquarters:  Madrid, Spain

Active since:  2011

 One original platform with a very good sense of the networking value. Talentous provides an cool design for your profile.

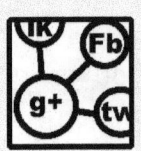 You can create your professional profile and make it public for other talentous users to evaluate and tip you on how to improve your presentation contact

 Although the website is only in Spanish, the concept is really good and invites for the interaction between members not only for business reasons but also for the networking spirit itself.

# RESUMUP

**9**

Web address:     **www.resumup.com**

Headquarters:             St.Petersburg Russia

Active since:             2011

The Russian Resumup, it was selected as Start up of the year in 2012 by Wired Magazine. A great tool for HR professionals.

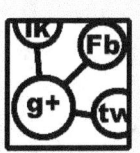

Your profile is ready in less than 10 seconds when you interface it with a Linkedin account and you can export it to PDF. HR managers can review a candidate's experience easily and they can compare it with other ones.

In an internet overloaded with professional networks resumeup.com shows up for its minimalistic style and produces some stunning info graphics of your professional career

# DOYOUBUZZ 10

Web address:        **www.doyoubuzz.com**

Headquarters:            Nantes France

Active since:              2010

 If online media buzz is your thing, then you are looking for Doyoubuzz.com. This is a French production that shows good quality and efficiency.

 It offers an incredible solution to synchronize all your status updates for all your network platforms and also improves your appearance in search results using (SEO).

 A smart presentation that makes it very easy to use also it has good filters to set up your privacy standards.

18

# MODELO

Web address: **www.modelocurriculum.net**

Headquarters:          Spain

Active since:          2008

You will not be impressed at first sight, and even if you compare it to other networking platforms, modelocurriculum do not offers more or better CV templates designs.

But what really will struck you is the amount of blogs, update and members interaction and blogs. There is no link with any other network so you will have to create from scratch your CV.

The main page is in Spanish but you have other European languages available, also you can find English articles as well as German updates on the job market.

# CUVITT

Web address:                    **www.cuvitt.com**

Headquarters:                   Madrid, Spain

Active since:                   2011

 Lets put it in few words, Cuvitt is a good idea. With more than 30 designs templates to create your resume it also offers an 'intelligent' analysis to understand your position in the market.

 You can interface your profile with your Linkedin or facebook account and you can also export it to PDF.

 Cuvitt is a different offer in the market and you are going to need over 30 minutes to input all your information to create your profile, but it worth the time.

# CVAUDERE 13

Web address:          **www.cvaudere.com**

Headquarters:          Spain

Active since:          2012

 100% in Spanish offers a good solution for those Spanish speakers to create a CV online using one of its many design templates.

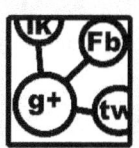 The presentation is clean and simple and also you can export your final production to PDF. It offers a good variety of templates to create a CV.

 The website is only in Spanish but it is worth to check it out. It is a very user-friendly site, an ideal solution for those who never did an online project or have little knowledge on Computers.

21

# PREZI

Web address: **www.prezi.com**

Headquarters: San Francisco USA

Active since: 2012

 Prezi uses Adobe Flash to create dynamic presentations in a very easy and smart way. By far one of the most innovative offers online.

 It is one of those 'out of the box' concepts, where instead of creating your presentation in different pages, you can do it in just one big dynamic presentation 'map' with zoom and pan effects.

 Give it a try, it might be a little difficult to figure out how to organize the ideas at beginning but after few minutes your will love it.

**RECOMMENDED!**

# MSOFFICE

Web address: **www.office.microsoft.com**

Headquarters:                USA

Active since:                1989

 The 2013 MS Office is a big step forward since it 2008 version or 2010 Mac edition. You can find some of the new templates online.

 Not only for the different template setups and formatting solutions to create your CV but also because now you can go online and get more out of the program.

 It is todays universally accepted format, so it might be a good option for those looking for no hassles nor complications.

**RECOMMENDED!**

# VIZUALIZE.ME 16

Web address:   **www.vizualize.me**

Headquarters:   Toronto Canada

Active since:   2011

 An excellent platform to create CVs with an info graphic design. Designed for the 'no designers'. They won back in 2011 the Toronto Start-up Weekend, and since then they are a non stop growing network.

 Vizualize.me shows a good example on the many ways you can promote your professional profile. With one click you can import all your data from Linkedin.

 And one example of that is 29.99U$ resume printed t-shirt offer online. Something different no doubts.

# PDFCV

| | |
|---|---|
| Web address: | **www.pdfcv.com** |
| Headquarters: | London, UK |
| Active since: | 2012 |

 The brain child of Lithuanian Software engineer Tadas Tamošauskas. This website is an efficient assistant for the creation of your professional online CV and for the conversion of the

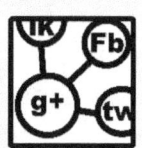 The presentation is simple and clear on how to use it. Also, allows you to register and log in using different social media platforms such as Facebook or Google +.

 It is a very easy-to-use site, all in English and an ideal solution for those who never did an online project and need some automatic assistance.

# KINZAA

**18**

Web address:          **www.kinzaa.com**

Headquarters:        California

Active since:          2011

 The info graphics are becoming the market standard for the resume presentation. More and more HR executives are using online CV networks to headhunt potential candidates.

 The platform has an interface with Linkedin so it makes it easy to import your data. Also you can set up your log in account using facebook.

 Here with Kinzaa you have the possibility to easily set up and create your online professional CV and make available for potential employers. You can even upload your own video

# COMOTO

Web address :        **www.comoto.com**

Headquarters:        Denmark

Active since:        2011

Comoto.com is a Danish production with the Scandinavian simplicity and design. You can import all your data from Linkedin to create your profile.

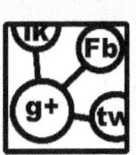
You can also export your final production into .PDF- .DOC or .PNG. The site is available in English, Danish and 5 other European languages.

For those looking for a simple and quick way to create your professional profile online Comoto.com is the place to do it. Some of the unique features Comoto offers: the possibility to create multiple Cvs.

**RECOMMENDED!**

# SLIDE ROCKET

Web address:   **www.sliderocket.com**

Headquarters:   San Francisco USA

Active since:   2007

 One of the best tools to create presentations. It makes cooperation on a single document very easy keeping the same communication standards for everybody.

 You can register and log in using different social media platforms such as Facebook or Google +. You can admin a group and customize the presentation.

 The idea of having one single space where you can zoom in and out to pin point the different parts of your presentation makes it a 'out of the box' idea when it come to online group presentation solutions.

# EASY-CV

Web address:        **www.easy-cv.com**

Headquarters:        France

Active since:        2008

 With easy-cv you can create a link so potential employers can access your profile online. The content is in French Spanish and other European Languages.

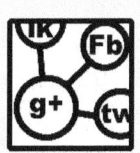 You have really good privacy filters and it do not requires of technical knowledge to work yourself around the page settings. There are no interfaces to set up your login account with facebook or LinkedIn.

 The Its all about the advantage of uploading not only the information of your career and skills but also you can add videos, photos or audios.

# VIZIFY

Web address:                 **www.vizify.com**

Headquarters:                Oregon USA

Active since:                2011

Many compares Vizify with about.me, but this network platform has a wider view when it comes to profile display. Founded in Portland Oregon by 3 young entrepreneurs who understood what was missing in the market.

Really easy to set up and operate. You can sign in by using Linked-in, Facebook, twitter , foursquare or even an instagram account, but if you don't have any of those, well you will need one.

You can even see some influence of Prezi.com at first sight, but it is under any circumstance a substitute for any of the mentioned websites.

# RESUMESIMO 23

Web address:  **www.resumesimo.com**

Headquarters:  Czech Republic

Active since:  2010

 Resumesimo is simple, clean, and easy to use platform for those looking for a quick way to have a CV online. The idea became really popular and from local moved to global. The company is based in Prague.

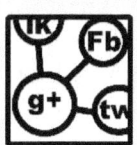 It interfaces with many other platforms and also offers to link your login information with other social networks, such as Facebook or Twitter.

 In only few steps you can create your professional online resume and export it to PDF for your own personal use.

# EXPRESS-CV

Web address:        **www.express-cv.com**

Headquarters:          Spain

Active since:          2010

 The Standard CV is based on the Europass System and in just 3 single steps you can create an high quality professional resume for the European market.

 Although there is no available interface with Linkedin or any other professional platform, the process is so easy and quick that really don't matter.

 The website is in Spanish with no other language option. Each section has an automatic assistant that will guide you step by step to fill up the forms with your personal and professional information.

# ONLINE PROFESSIONAL NETWORKS

www.linkedin.com

www.viadeo.com

www.xing.com

www.youngenterpreneurs.com

www.ziggs.com

www.sunzu.com

www.cofoundr.com

www.biznik.com

www.proscore.com

www.efactor.com

www.spoke.com

www.meetups.com

www.data.com

www.upspring.com

www.perfectbusiness.com

www.startupnation.com

www.talkbiznow.com

www.younoodle.com

www.gust.com

www.fledgewing.com

www.ushi.com

www.onlinemarketinglatan.com

www.guru.com

www.jingwei.com

www.grera.com

www.plaxo.com

# LINKEDIN

Web address:                **www.linkedin.com**

Headquarters:               California USA

Active since:               2002

 Linkedin became in a short time the 'must have' professional profile with more than 200 million registered members. About half of the members are in the United States and 11 million are from Europe.

 It is important to have your Linkedin profile for many reasons, one of that most of the new professional platform will offer you to interface your Linkedin profile into their own website.

 It is really easy to use and provides on of the best customer services in the market. It worths every penny.

# VIADEO

**26**

Web address: **www.viadeo.com**

Headquarters: Paris France

Active since: 2004

Viadeo is the second largest social and professional network, active since 2004 and created by Dan Serfaty and Thierry Lunati. Viadeo lets members maintain a list of business partners, allowing them to stay in touch, use or help each other to find a job, or create business opportunities.

This is a very strong platform in France and it has expanded from Europe to the EMEA region within time, reaching over 50 millions members.

The website is set up in several European languages and shows a high blog activity from its members. To sign in is a simple step.

# XING

**27**

Web address: **www.xing.com**

Headquarters: Hamburg Germany

Active since: 2003

With its 11 million users XING became one of the most popular professional platforms in Europe. Basic membership is free. XING also offers the system for closed communities, called Enterprise groups.

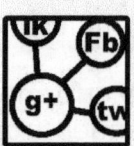

Having your profile with XING will not conflict with your profile on Linkedin. It is in fact a more personalize solution than LinkedIn and being a European network helps to connect with other local HR platforms.

**RECOMMENDED!**

The display is clean a simple and it is really easy to use. It will take few steps to create your profile.

# YE

**28**

Web address:
**www.youngenterpreneurs.com**

Headquarters: USA

Active since: 1999

 Better known as YE, Young Entrepreneurs is the network for young university student with promising projects and looking for an angel investor, mentoring or advice. Todays is one of the largest online forums for entrepeneurs.

**RECOMMENDED!**

 Many head-hunters wonder around this network in search of potential candidates for top executive positions.

 You must read the forums, one of the most active features of the website. It is very easy to become a member just check it out and you will see.

# ZIGGS

Web address: **www.ziggs.com**

Headquarters: California

Active since: 2010

 Like Linkedin and XING, Ziggs offers a platform easy to use and helping you to create your own professional self 'brand', controlling, top visibility in search engines.

 Ziggs will allow you planning and executing an effective marketing strategy to promote your professional profile online in connection with other networks. There is no interface with other networks to set up your log in.

 It is worth to give it a try and test the benefits of Ziggs solutions.

# SUNZU

Web address: **www.sunzu.com**

Headquarters: Dublin, Ireland

Active since: 2013

 Created in 2013 from the e-cademy and based in Dublin city, Sunzu is one of the post popular sites in Europe providing a fantastic combination of Social and Professional networking.

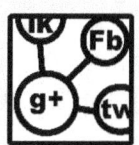 Statistic, Status updates, tags, articles, blogs and many other application are available depending on the package you choose. No interface with facebook or Linkedin available for setting up your log in.

 No doubts that Sunzu is the Apple of professional networks. It just works.

# COFOUNDR 31

Web address:                **www.cofoundr.com**

Headquarters:               USA

Active since:               2012

 Cofoundr is cool, and that why you must try it. Even if you are searching for a job, the networking dynamics of this site makes it worthy to be part of.

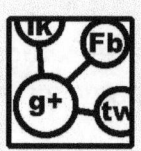 It is possible to interface your Linkedin directly with your Cofoundr account. It also offers to synchronize your login with Facebook.

 Mainly you will find entrepreneurs looking for a start-up funding, you will also find new ventures looking for executives for their initial stage.

40

# BIZNIK 32

Web address: **www.biznik.com**

Headquarters: USA

Active since: 2005

 Biznik is the perfect space for those having their own business and want to dedicate time and energy to develop online networking and promotion.

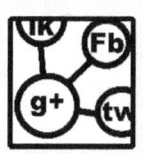 It is not just about adding contact into your profile but the sites promotes users to meet up face to face making the network itself stronger and real.

 For those looking for a simple and quick way to create your professional profile online this is the place to do it.

**RECOMMENDED!**

41

# PROSKORE 33

Web address: **www.proskore.com**

Headquarters: Florida USA

Active since: 2012

 Imagine that your profile can be scored based on your experience, amount of contacts, online media buzz and activity. PROskore is an online business network that measures professional reputation

 Very easy to set up and you can log in using your social network credentials (eg: Facebook). PROskore analyze each professional background (i.e. education, work experience) along with their social media influence (followers, engagement, etc.).

 Make sure you review the privacy settings before you start your ProScore experience.

# EFACTOR  34

Web address: **www.efactor.com**

Headquarters: California USA

Active since: 2008

 In March 2008 Efactor.com started its activity based on the Netherlands. By July 2011 it reached its 1$^{st}$ million members from 185 countries.

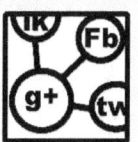 Most of the users are from the UK the US and Holland. Setting up your account takes a little time but its worthy to go thru it. No interface available with LinkedIn or facebook to set up your account.

 The display is clean a simple and it is really easy to use. It will take few steps to create your profile.

# MEETUP

Web address:            **www.meetup.com**

Headquarters:           New York USA

Active since:           2001

 Bringing the whole online network experience to the life of a neighbourhood or community, that is what meetups.com is about.

 A network for clubs, associations, and social communities wanting to connect with each other for cooperation in projects, events or activities.

 A very easy to use website and with a relaxed and informal presentation. Really worth to check what is about. Users enter their ZIP code or their city and the topic they want to meet about, and the website helps them arrange a place and time to meet.

# SPOKE

Web address:       **www.spoke.com**

Headquarters:       California

Active since:       2003

 It offers to a space for companies, sales executives, HR and individuals looking to expand their professional network. co-founded by Ben T. Smith in 2002.

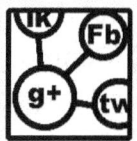 You can set up your log in setting by link any of your social platform (Facebook, twitter or Google+).
They have a good database to search for potential clients.

 Each section has an automatic assistant that will guide you step by step to fill up the forms with your personal and professional information.

45

# DATA

**37**

Web address: **www.data.com**

Headquarters: California USA

Active since: 2012

What before was jigsaw.com is now data.com, property of salesforce, the company dedicated to CRM solutions.

With over 30 million business contacts is a promising venture that will soon to be a networking boom. Data.com Connect's unique Wikipedia-style crowd-sourcing model delivers the world's most complete, accurate and up-to-date business contact and company data.

A simple concept that is creative too. It sums up the experience by sharing your profile as a business card.

# UPSPRING

Web address: **www.upspring.com**

Headquarters: USA

Active since: 2012

 For the small guy upspring provides a fair space. One original platform with a very good sense of the networking value.

 Small and Medium companies and can relate to this website to contact and expand their professional networks.
You can log in using Facebook, twitter or your email account.

 With more than just a profile, upspring offers a matching system filtering industries, professions, products or services.

# PERFECT BUSINESS

# 39

Web address:     **www.perfectbusiness.com**

Headquarters:              California

Active since:               2011

 Entrepreneurs and Executives can equally use this platform in order to develop their online professional profiles and expand their networks. PerfectBusiness also provides professional business planning software, startup resources and inspiring interviews with leading entrepreneurs

**RECOMMENDED!**

 Blogs, courses and updates from different industries and projects can be found in perfectbusiness.com.

 It is a very easy-to-use site, an ideal solution for those who never did an online project and need some automatic assistance.

# STARTUP
# NATION

Web address :        **www.startupnation.com**

Headquarters:              USA

Active since:              2002

It is a site for start ups yes, but it gathers all the right contacts you will need to expand and develop your online professional network. Jeff and Rich Sloan are company creators, lifelong entrepreneurs, and brothers.

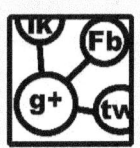

Think about as a directory of future potential employers. You can register and log in using different social media platforms such as Facebook or Google +.

It is really easy to set up your account and you can access immediately to other members profiles.

# TALK BIZ
# NOW

Web address:        **www.talkbiznow.com**

Headquarters:        California USA

Active since:        2008

Ok, imagine having a control panel with all the information of all your social and professional networks online.

Talkbiznow can enhance your productivity online by effectively centralizing all your data into one page.

You can even work out an option to create webinars online for any given group of contacts. By far one of the most effective set up and displays in the offer now a day.

# YOUNOODLE 42

Web address:          **www.younoodle.com**

Headquarters:              California USA

Active since:              2007

 Visual wise Younoodle is like the instagram of professional networks. The aesthetics are very similar as well the simplicity and smart design. The founders are Bob Goodson and Kirill Makharinsky, from University of Oxford.

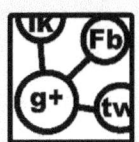 You can register and log in using different social media platforms such as Facebook or Google +.

 YouNoodle's "Startup predictor", developed by Makharinsky and Hwang, uses mathematical models to predict the success of new businesses.

51

# GUST 43

Web address: **www.gust.com**

Headquarters: New York USA

Active since: 2004 (angelsoft)

 With over 160.000 start-ups Gust aims those who seek potential partners in projects. Gust presents a great design and easy to use format for those starting up a project and wants to get around the professional networking.

 Founded by David S. Rose He was named to the Inc 500 list as CEO of one of the country's fastest growing private companies, and has been described as a 'world conquering entrepreneur' by BusinessWeek.

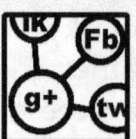 You can register and log in using different social media platforms such as Facebook or Google +.

# FLEDGE WING

# 44

Web address:     **www.fledgewing.com**

Headquarters:     USA

Active since:     2009

 An online community focusing on University undergrads and young entrepreneurs, launched on February 2009 by two students in NYU.

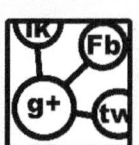 It expanded so far to 170 universities worldwide and it's a promising platform to become a top leader for the young start-up leaders connecting them with mentors, angel investors and advisors.

 It is one of the promising platforms that for the past few years added many members and it is still growing.

# USHI

**45**

Web address:               **www.ushi.com**

Headquarters:              Shanghai China

Active since:              2010

 For those willing to venture into new territories Ushi.com is the place to be. Ushi makes it easier to find new customers, partners, employees, jobs, and experts, and enables members to share and discuss information.

 This professional online network from China puts together C-Level and top management executives from all over Asia.

 You will have to work out the linguistic barrier with google translator, not easy, but not impossible. It is indeed worth to investigate and make your own judgement on what you can learn from it.

# ONLINE MARKETING 46

Web address:
**www.onlinemarketinglatam.com**
Headquarters:               Argentina

Active since:               2009

 In a slow but steady way, more online professional networks are appearing in Latin America. Check out this Argentinian hub.

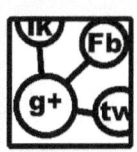 This professional online network from Argentina puts together C-Level and top management executives from all over South America.

 You will have to work out the linguistic barrier with google translator, not easy, but not impossible. It is indeed worth to investigate and make your own judgement on what you can learn from it.

# GURU

**47**

| | |
|---|---|
| Web address: | **www.guru.com** |
| Headquarters: | San Francisco USA |
| Active since: | 1998 |

 A very well organized professional network for freelancer and independent projects. One of the largest online market place for online talents.

 There is no direct interface between guru.com and other professional networks. So you will have to type all your profile details.

 You can register as an employer or a freelancer. There is a lot of good information in the blog section. Check out the online tour to discover all the cool features.

**RECOMMENDED!**

# JINGWEI

Web address:                    **www.jingwei.com**

Headquarters:               Chinese

Active since:                2004

 One of the most popular websites in China, created by the founders of Zhopin.com, the website with the biggest amount of job offers in the world.

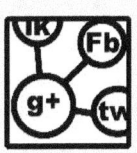 Jingwei.com it is not the traditional professional website, it looks like a combination of LinkedIn and Quora, that generates an interaction between members in a unique fashion.

 The website is only available in Chinese but sure you will find your way around using google translator. Anyway it is worthy to check it out.

# GRERA
# 49

Web address:             **www.grera.net**

Headquarters:        Barcelona, Spain.

Active since:         2010

 The perfect combination of classified adds and networking profiles. Grera is Spain top platform for executives from all industries to meet and exchange information about offers and demands on products and services.

 The true value of this site is the automatic interface between small and middle size companies to perform combined purchasing with better pricing.

 For those willing to face the linguistic challenge (the site only is in Spanish) will find the most updated database in Spain.

# PLAXO

# 50

Web address: **www.plaxo.com**

Headquarters: California USA

Active since: 2002

 Plaxo is an online address book and social networking service originally founded by Sean Parker, Minh Nguyen and two Stanford engineering students. With Plaxo you can synchronize in real time all your agendas and calendars using mobile devices.

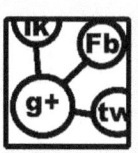 You can set up your log in using Facebook, google, yahoo o Hotmail.

 Plaxo concentrates all your contacts from all your social and professional networks into one screen and updates them when someone changes or updates the information.

# RECRUITMENT AGENCIES

www.michaelpage.com
www.prospects.com
www.adecco.com
www.manpower.com
www.globalcareer.com
www.antal.com
www.mercuriuval.com
www.pagepersonnel.com
www.europeanrecruitment.com
www.theladders.com
www.europeansolutions.nl
www.emearecruitment.com
www.hays.com
www.tate.co.uk
www.carlisle.co.uk
www.ramstad.co.uk
www.russelreynolds.com
www.hendrikstruggles.com
www.kornferry.com
www.spencerstuart.com
www.egonzhender.com
www.approachpeople.com
www.aims-international.com
www.teamwork-one.com
www.cgconsultants.com
www.careers-global.com
www.europrojects.com

# MICHAEL PAGE

**51**

Web address: **www.michaelpage.com**

Headquarters: Surrey UK

Active since: 1976

Michael Page is the big brother of all headhunting agencies. The Company was formed in 1976 by Michael Page and Bill McGregor. In 2006 Steve Ingham was appointed CEO.

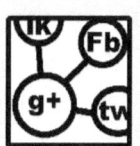

With offices all around the world they are with no doubt the biggest and most organized HR network, working with the top private and public enterprises. You can join them on Linkedin.

Page International renamed as Page Group started operating under three different brands: Page Executive, Michael Page International and Page Personnel.

# PROSPECTS

Web address: **www.prospects.com**

Headquarters: Manchester UK

Active since: 2009

 Prospects focus on the online development for it professional network and job offers. They are a strong company with a good reputation in the UK and Europe.

 Numbers are the best evidence of how successful Prospect is becoming. Over 1 million unique visits and more than 6 million clicks by October 2011.

 You can access immediately to job offers by setting up your account in a easy and quick manner. You can also follow them in popular networks such as Facebook, Linkedin or twitter.

**RECOMMENDED!**

# ADECCO

**53**

Web address: **www.adecco.com**

Headquarters: Glattbrugg Switzerland

Active since: 1996

Adecco is recruitment company with head offices based in Switzerland with more than 28000 employees in its 5500 offices in over 60 countries and territories around the world.

Being one of the worldwide known HR company Adecco hires over half million jobs seekers a year. it offers a wide variety of services, connecting over 700,000 associates with well over 100,000 clients every day.

You can choose your local Adecco office to start your search. Check out the blog, it is really good and it has valuable information.

# MANPOWER 54

Web address: **www.manpower.com**

Headquarters: Wisconsin USA

Active since: 1948

 For over 60 years Manpower has dedicated to enrich the life of people with a trustable recruitment process and ensuring equal opportunities in the job market. Their customer base, is approximately 400,000 companies each year.

 You can tastes their innovative spirit thru their website where you can locate and contact directly a local representative. As of 2011, Manpower Group has over 3,900 offices in 82 countries.

 It is really easy to set up your account and you can access immediately to job offers.

# GLOBAL
# CAREER

**55**

Web address:
**www.globalcareercompany.com**
Headquarters: London UK

Active since: 2002

After 10 years of solid development in the recruitment industry Global Career is todays one of the top leaders for the executive search

With a solid network in the emerging markets working with companies from North Africa, Asian Pacific and India, Brazil and CIS region. They have recruited internationally for over 400 companies across the globe.

It is really easy to set up your account and you can access immediately to job offers.

# ANTAL

Web address: **www.antal.com**

Headquarters: London UK

Active since: 1993

 With an organization structure that covers over 33 countries with 107 offices Antal International Ltd is one of the top ten recruitment agencies to be consider.

 Their headquarters are based in the UK but you can access in their website to the contact information of their network across the world.

 Very innovative 'new job offer' feeds displayed on the home page. What really calls the attention from other recruitment agencies is that Antal has a Charity foundation, check it out.

# MERCURI URVAL

**57**

Web address:     **www.mercuriurval.com**

Headquarters:     Sweden

Active since:     1967

Mercuri Urval International AB is the Scandinavian leader in recruitment and executive search.

Based in Sweden and launched in 1967 its one of the most reputable and solid HR company in northern Europe. Today the company has expended to over 25 countries with more than 700 staff members around the world.

You can choose your location on the main page and job offers will be display according to your selected filters. Applying is really simple, you just need to upload you PDF file.

# PAGE PERSONNEL 58

Web address: **www.pagepersonnel.com**

Headquarters: Sweden

Active since: 1994

 Page Personnel operated in 21 countries around the world. They are a subsidiary of the Page Group. The Page Group is a FTSE 250 constituent with over 35 years of experience in professional services recruitment.

 Based in Sweden and launched in 1967 its one of the most reputable and solid HR company in northern Europe.

 They have a very good display of job offers in their website, where you can register and add your self to a mailing list for monthly notifications about job offers and market

# EUROPEAN RECRUITMENT 59

Web address:           **www.eu-recruit.com**

Headquarters:          London UK

Active since:          2009

 They specialized in executive search for Software developers, Hardware Engineers Wi-Fi specialists and many other top professions form the IT, media and communications industry.

 Founded in 2009 by David Wicks and rapidly expanded it business across the UK and Europe. openings.

 They have an interesting display on their home page making it very attractive to the job seekers to register and search for job openings.

69

# THE LADDERS 60

Web address:  **www.theladders.com**

Headquarters:  New York USA

Active since:  2003

Based in the USA The Ladders provides an online service for job search. With most of their jobs offer pointing into the top executive segment, The Ladders was one of the first one to publicly start offering 6 figures jobs online.

The website is organized in a simple manner with no distractions other than the job search filters. Register your profile and receive weekly email updates with job offers.

They offer a basic membership for free that will give you a good picture on the benefits of paying for a Premium membership. There is an special log in for employers.

# EUROPEAN SOLUTIONS

**61**

Web address :   **www.europeansolutions.nl**

Headquarters:        Netherlands

Active since:        2006

Based in the Netherlands but with a wide network across the EMEA region , European Solution is leader in the market for recruitment and HR development for the public and private sector.

As a difference from other Agencies European Management Solution also works with job offers for mid-career and senior specialists, as well as free-lance consultants, making it the top leader in the HR industry for this kind of missions.

It is really easy to set up your account and you can access immediately to job offers.

71

# EMEA RECRUITMENT 62

Web address:
**www.emearecruitment.com**
Headquarters: California

Active since: 2011

 EMEA Recruitment was founded in 2007 by Paul and Kelly Toms EMEA Recruitment are specialists in recruiting top executives for HR, Accounting and Finance position across the EMEA region.

 Their offices are in Vienna, Zug and Amsterdam and they have a wide network of executives head hunters across the EMEA region managing top accounts for executive search.

 The home page displays pretty much all the info at first glimpse. Its easy to create an account. They offer a 415Euros if you refer a candidate and it is successfully hired.

# HAYS

**63**

Web address:       **www.hays.com**

Headquarters:      London UK

Active since:      1867

Hays PLC is a British company dedicated to recruitment and HR consultancy. They are valued at the London Stock Market and it is rated in FT250.

They work out projects across Europe, Asia and the UK.
Subscribe today to their websites and you will be able to access a wide offer of top executive job openings.

Check out the Global Skills Index, a really good report for your job hunting. Just check your location from the Hays worldwide drop list and you are ready to

# TATE

**64**

Web address: **www.tate.co.uk**

Headquarters: California

Active since: 1985

Since then Tate Recruitment agency became one of the most reputable and popular agencies in the UK managing a wide arrange of accounts across the region and country.

Mrs Virginia Tate, who aimed to create a boutique recruitment agency, founded Tate back in 1985. An award winning recruitment agency in the UK.

The home page already offers you filters to start your search. You can also download podcasts and follow them on the major social media channels.

# CARLISLE MANAGEMENT **65**

Web address: **www.carlislems.co.uk**

Headquarters: London UK

Active since: 1997

 Carlisle Management Solutions was born out of Tate recruitment agency in 1997 by Gill Stewart. They hold an ISO 9001 Quality Management Standard.

 With more than 80 staff members Carlisle team manage top accounts in the region and keeps a wide variety of job offers posted on their website.

 Register with them for more information and updates on the UK market. Email and phone number are available on the home page.
Check their impressive client list.

# RANDSTAD 66

Web address: **www.randstad.co.uk**

Headquarters: Netherlands

Active since: 1960

 Randstad was founded in 1960 by Frits Goldschmeding.Ramstad Holding is an international recruitment agency, which focuses on temporary personnel and HR administration for top

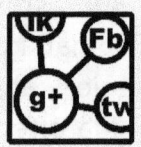 Every day over half million temporary workers are hired thru Randstad. Around the world Ramstad operates with over 3000 agencies.

 No doubts that it is a must to stop by their website and have a look around. select your industry sector or expertise in the drop down menu and upload your CV for the selected job.

# RUSSELL REYNOLDS

Web address :    **www.russellreynolds.com**

Headquarters:           New York USA

Active since:            1969

 Russell Reynolds Associates is one of the top leaders in executive search. You can locate your local Russell Reynolds Representative in 41 offices around the world.

 It is no doubts the right option for C-Level and VP looking for the next move n their careers. Check the video series section.

 Russell Reynolds website is easy to use and shows all contact details of its associates. The website is in English and Chinese. And you can have direct access to the Russell Reynolds consultants directory.

77

# HENDRICK & STRUGGLES

# 68

Web address: **www.hendrickstruggles.com**

Headquarters:                    Chicago USA

Active since:                    1953

 Heidrick & Struggles International is a HR company that specialize on the recruitment of top executives around the world.

 The company offers a wide variety of services besides recruitment and HR administration. They became well known since they hired Eric Schimdt Google CEO back in 2004.

 You will find all the information you need about the company and how to search and apply for a job in the home page. You can look up a consultant by industry or location.

# KORN FERRY

**69**

Web address: **www.kornferry.com**

Headquarters: Los Angeles USA

Active since: 1969

Korn/Ferry is one of the most influential executive search agencies in the world. They have over 80 offices and nearly 3000 employees managing job offers from top executive firms.

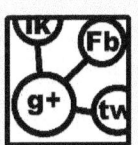

Their home page is clean and clear to read, you will have access to all the public job offers. Make sure to register your profile and to get in contact directly with on of their agents.

The home page is simple in design. You can search for any consultant by location, name or industry. Follow them in any of the major social media channels.

# SPENCER STUART

 70

Web address:     **www.spencerstuart.com**

Headquarters:          Chicago USA

Active since:          1956

 Active since 1956 Spencer Stuart is a private company specializing in executive and c-level search consulting for top companies around the world.

 They operate with 54 offices in 29 countries performing over 4000 contracts a year across the world. They work with more than 300 consultants, making their network on of the biggest ones.

 You can register on their website and also receive email updates on job offers and market news. Don't miss the' Research and Insight' section with great market reviews.

# EGON ZHENDER

**71**

Web address:     **www.egonzhender.com**

Headquarters:          Zurich Switzerland

Active since:          1964

Being today one of the most trustable Agencies by top companies in Europe, Egon Zhender works with more than 420 recruitment consultants based on 66 offices in 40 countries around the world.

This fast growing company is becoming the favourite for big companies looking for top executives. They work with the top companies worldwide.

The FOCUS Magazine will give you all the information you need for market updates and the current trends on the C-Level management.

# APPROACH PEOPLE

Web address: **www.approachpaople.com**

Headquarters: Dublin Ireland

Active since: 2000

 Approach People Recruitment is a young and dynamic company that opened doors in 2000 and since then they became a reference in the international recruitment.

 By far they are one of the innovative ones when it come to online display of offers, something you can appreciate at first sight when you open their homepage.

 Choose from the location filters in the main page to start searching for a job. Check the newsletter in the candidate section.

# AIMS INTERNATIONAL 73

Web address : **www.aims-international.com**

Headquarters:                 Vienna Austria

Active since:                 1993

 A one of a kind agency, AIMS International operates since 2004 with more than 90 offices in 50 countries across EMEA and APAC. In 2006 they have been nominated as the biggest HR network in the world with over 350 consultants.

 You can find your local representative in their website, also you can access the updated job offers without having to register your profile.

 You can register on their website and also receive email updates on job offers and market news.

# TEAMWORK ONE

**74**

Web address : **www.teamwork-one.com**

Headquarters: Germany

Active since: 1998

Team Work One accomplished one of the most complete databases in the hospitality and restaurant industry thanks to their experience recruitment consultants.

That know-how is the main reason why major European companies choose Teamwork one for their next executive search.

The website is easy to follow and you can start your seach on the main page by choose the options from the filters menu. Applying to the selected job takes few seconds.

# CG CONSULTANTS 75

Web address : **www.cgconsultants.com**

Headquarters: UK

Active since: 2008

 Based on the United Kingdom CG Consultants is a fast growing recruitment and executive search firm operating at a national and international level since 1976.

 They specialize mainly in the scientific sector offering permanent positions. They hold a very good database of candidates for outsourcing solutions and permanent positions.

 The home page is simple and straight to the point. You can send an enquiry on their website.

# CAREERS
# GLOBAL

**76**

Web address: **www.careers-global.com**

Headquarters: India

Active since: 1999

Launched in 1999, Careers-Global is one of the top agencies in online recruitment in India

You can subscribe to their network by submitting your profile information on their website, you will be able to receive email in a weekly period with information about job offers and market updates.

Click on the 'Latest Availability' option form the main page to get thru the job offers list or you can just submit your resume by clicking the 'submit resume' option

# EURO PROJECTS

Web address:      **www.europrojects.co.uk**

Headquarters:              London UK

Active since:              1996

 Launched in 1996 as a specialized service for headhunting and recruitment for engineers, technical executives and scientist.

 Today Euro Projects experienced team reaches a network across Europe and the EMEA region. The Client Resource Centre is a dedicated area providing essential information on the hiring process and advice.

 They have nice designed website and it is fairly easy to set up an account with them. Don't forget to check the resource centre.

**RECOMMENDED!**

# ONLINE JOB BOARDS

www.6figurejobs.com
www.exec-appointments.com
www.indeed.com
www.expeteer.com
www.monster.com
www.stepstone.com
www.snagajob.com
www.ceevee.com
www.simplyhired.com
www.naukrigulf.com
www.boyden.uk.com
www.cwjobs.co.uk
www.executivesontheweb.com
www.totaljobs.com
www.barnabystewart.com
www.careerbuilder.com
www.glassdoor.com
www.reed.co.uk
www.cvtrumpet.co.uk
www.eurojobs.com
www.idealist.com
www.fish4jobs.co.uk

# 6FIGURE JOBS

**78**

Web address: **www.6figurejobs.com**

Headquarters: UK

Active since: 1999

 More than 700,00 top executives and professionals are registered at 6figuresJobs to promote their profile among potential employers from F500 companies around the world.

 6FigureJobs only offers $ 100K + job announcements for C-level, VP and Directors jobs worldwide. There are 650,000+ pre-screened executive and senior-level job seekers.

 You can find on 6FigureJobs webpage career and salary advice. There is a basic membership option free of charge available to job hunters.

# EXEC-APPOINTMENT 79

Web address:
**www.exec-appointments.com**
Headquarters:                London UK

Active since:                1999

Exec-appointments LTD is a subsidiary of Financial Times, by far the best online job board for C-Level and VP careers search. With offices in the UK, New York and Dubai. They are the leading global job board for both the private and public sectors.

**RECOMMENDED!**

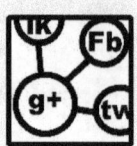

The website is easy to use and the have a free of charge email notification service. But no interface with Linkedin or Facebook.

The search filter options are available straight in the front page. An interesting feature is the recruiters list A to Z that publish on exec-appointments.

90

# INDEED

**80**

| | |
|---|---|
| Web address: | **www.indeed.com** |
| Headquarters: | USA |
| Active since: | 2004 |

Indeed.com the job searcher with more job offers in the US online job market. It is available in 53 countries and in 2010 Indeed.com surpassed Monster.com as the most visited in the United States.

Actually holds 60 millions unique visits every month. It is by far a clear example of how online job websites should be in design and functionality.

A very 'Google' style home page, just type location and industry and you can start searching for a job. The website is available in over 10 languages.

# EXPETEER

Web address: **www.expeteer.co.uk**

Headquarters: London UK

Active since: 2007

 In this platform you can find job offers and HR teams from private companies. What really catches your eye about this site is the aesthetics and home page display, making it very easy to use.

 Experteer offers a subscription without a fee and you can register your profile to receive weekly jobs notifications. There is no interface with Linkedin or facebook to log in.

 With a selection of more than 90,000 job offers you can start searching for your preferred job using the main page filters for location, salary and industry.

# MONSTER

Web address: **www.monster.com**

Headquarters: New York USA

Active since: 1999

Worlds number one online job board Monster.com owned and manage by por Monster Worldwide, Inc. The company has over 5000 staff members in 36 countries across the world .

In 2006 Monster.com was one of the 20th most visits websites in the world, with over 100 million visits (according to comscore media metrics).

Really easy to manage and search for jobs. Check out the top questions section, you will find tons of useful information.

93

# STEPSTONE 83

Web address: **www.stepstone.com**

Headquarters: Berlin Germany

Active since: 1996

 Active since 1996 StepStone is one the most visited European online job boards, based in the United Kingdom. They advertise over 11.2 miilion active subscribers.

 890 people works at Stepstone in their offices across 9 countries, putting together small and middle size companies with potential candidates.

 They have a clean and clear website and it is fairly easy to set up an account with them. The websites is easy to use and counts with email notification service.

# SNAGAJOB

Web address:                **www.snagajob.com**

Headquarters:              Richmond USA

Active since:                1999

 Snagajob was launched on October 17, 1999, by former attorney and current CEO Shawn Boyer. It is not the biggest one but is starting to be one of the most popular ones in the US job search market.

 The platform is very easy to use an you can subscribe to post your resume with preference filters by industry, profession or location.

 The website is fairly easy to set up an account with them. Check the right-fit degrees section. The websites is easy to use and counts with email notification service.

95

# CEEVEE

Web address: **www.ceevee.com**

Headquarters: Romania

Active since: 2010

On CeeVee you can find a good overview on the job market statistics per region or industry. CeeVee offer access to 300.000 job seekers from East-European countries.

Here you can see how many job offers exists per city or country, the website is available in English, German, French and 4 other European languages.

They have a clean and clear website and it is fairly easy to set up an account with them. The websites is easy to use and counts with email notification service.

# SIMPLY HIRED

**86**

Web address :         **www.simplyhired.com**

Headquarters:          California USA

Active since:           2005

Simply Hired is a job search website where you can find update on job offers from several other websites.

Making Simply Hired a hub for thousands of online offers. They operate in 17 countries and their feeds includes private company jobs posts, press classifieds , websites job posts and Blogs.

It is easy to set up an account with them just login with Facebook. Subscribe and you will get email notification service.

# NAUKRIGULF 87

Web address:     **www.naukrigulf.com**

Headquarters:     India

Active since:     2008

 This job search site is becoming one of the most popular in Middle East and India. Very complete and full with useful information about the job market.

 The home page is clear and strait to the point on its offer and present very good search filters to locate your preference on Industry, Profession or Location. A great database for job seekers.

 Have a look and dedicate some time to get use to the listings, as not all of them have the hiring company details.

# BOYDEN

Web address :            **www.boyden.uk.com**

Headquarters:            London UK

Active since:            1946

With over 70 years of experience operating within 44 countries worldwide, Boyden is a privately owned, partner-run business offering Executive Search and Interim Management services across all functional business disciplines.

Keeps updated information on recruitment trends in the European market. There is no interface with any of the major social networks, so you have to input your info manually.

The website is only in English and the presentation of the home page is easy to read and takes practically nothing to register your profile. .

# CWJOBS

**89**

Web address: **www.cwjobs.co.uk**

Headquarters: London UK

Active since: 2006

 CWJobs.co.uk job board specialize on IT careers. Every month over 320000 potential candidates visits the website looking for an job offer.

 The service is free of charge and easy to use. You don't need to register to look after job offer. Still you can subscribe to get job offer via email.

 Search the alphabetical list of direct employers with current jobs posted on CWJobs. The Career advice section is really good.

# EXECUTIVES ON THE WEB

**90**

Web address:
**www.executivesontheweb.com**
Headquarters:              UK

Active since:              2001

 All the jobs you find posted on executivesontheweb.com online board are filtered by salaries £50k and up. Top European companies works with them.

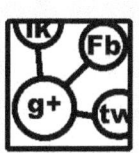 The service is free of charge and you can access the job offers by registering your profile details . You will not regret having weekly updates in you email from executivesontheweb.com.

 Pay attention to the media centre section, loads of information that will help you to get around the job ads.

# TOTALJOBS 91

Web address: **www.totaljobs.com**

Headquarters: London UK

Active since: 2003

 Totaljobs.com is one of the main job boards online worldwide and in the United Kingdom. The site has over 90000 job post every month attracting more than 3,6 million people looking for a job.

 Almost all industries in the UK uses Totaljobs.com to post their job offers. The website is very easy to use and you can register to receive email notifications about job offers.

 Register and make sure to set up the 'My Total Jobs' section and you will be able to receive weekly updates on the job market per location or industry.

# BARNABY STEWART

**92**

Web address : **www.barnabystewart.com**

Headquarters:                London UK

Active since:                1999

Barnaby Stewart Superior Executive Search & Selection operate in United Kingdom and Europe making them one of the most popular executive search companies in among the F500 companies.

More than 70% of their clients repeat business with them and their website set up shows the experience and quality of their job.

The presentation of the home page is easy to read and takes practically nothing to register your profile. .

103

# CAREER BUILDER

**93**

Web address: **www.careerbuilder.com**

Headquarters: USA

Active since: 1995

With 24 million unique visit every month CareerBuilder.com is the biggest online job board in the United States.
It operates in more than 21 countries with a presence in 55 markets.

The added value of this website is the partnership with key newspapers to promote their classifieds. You can set up your login using facebook or google +.

Look around the partners section and check the companies they work for/with. You can also do your job search by filtering per industry and location.

# GLASSDOOR 94

Web address:              **www.glassdoor.com**

Headquarters:             USA

Active since:             2008

 Glassdoor is an open source of the working community where the active users have the same view on the value of sharing salary information in top listed companies per country and regions.

 You can also fins feedback about companies and some internal information. The source became a really valuable piece in the HR and recruitment industry.

 With over 3 million anonymous reviews is a must check website. Don't miss the blog section with all the updates from the top managers in the key industries.

# REED

Web address: **www.reed.co.uk**

Headquarters: London UK

Active since: 1960

 Founded by Sir. Alec Reed in 1960 and manage by his son Mr. James Reed, today has over 3000 recruitment specialist in 180 countries around the world.

 They went online in 1995, and since then the company has grown as one of the leaders in the recruitment industry in the UK and Europe.

 You can connect with any of the major social networks. Make sure you go thru the 'In demand qualifications' section, it will give you a good overview of the market needs.

# CVTRUMPET 96

Web address: **www.cvtrumpet.co.uk**

Headquarters: London UK

Active since: 2002

 CVtrumpet has over 8200 agents receiving your CV once you register it. They claim that you will be reaching/covering 70% of all posted jobs in the region by subscribing with them.

 In few words it groups almost all major agencies contact worldwide and save you time by making your profile available for all of them in one single platform.

 The basic package is for free but read the packages section and read more about the membership fees and benefits .

# EUROJOBS

Web address: **www.eurojobs.com**

Headquarters: UK

Active since: 1993

 The granddad of all online job websites. Going back to 1993 it looked like an isolated action few had access too. But the leaders are made of innovators.

 So get your account today and search for a job that fits your profile. They have good filters to tip down what you are looking for. You can follow them in any of the major social network channels.

**RECOMMENDED!**

 Check out the blogs and subscribe for emailing alerts on jobs offers.

# IDEALIST

**98**

| | |
|---|---|
| Web address: | **www.idealist.com** |
| Headquarters: | USA |
| Active since: | 1995 |

If you are looking to get out of the rat race and start helping others with an altruist spirit than you must check idealist.

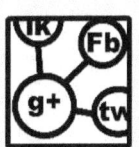

With offers to volunteer in projects all around the world it became on of the most popular sites for ONGs and Social Projects.

Make sure you read all about them and their vision, so you understand what is the company about and their mission.

# FISH4JOBS

Web address: **www.fish4jobs.co.uk**

Headquarters: UK

Active since: 2009

 Fish4jobs is one of the best known websites for job hunting in the UK. With 1.8 million registered candidates and over 1.3 million unique monthly visits fish4jobs offer job search results for different areas and industries.

 No doubts that you should register your profile and review the website result on jobs in the UK but also in Europe.

 Check the 'Advice' section for more information about activities eg: jobs road show in the orange double decker.

# SOME TIPS

Here some tips for your job search

1.- Consistency: Keep the same information in the profiles of all your networks.

2.- Diversify: Not all jobs are in one online job board. Select the ones of your interest and check them in a regular way.

3.- Google your name: That is the first thing an HR consultant will do in order to start a background check. Your life is online.

4.- Public Databases: Fortune 500 & Forbes 100 Are really good database for C-Level names and key people in top companies, but they are not the only ones. Check Finance Blogs and start up articles.

5.- Create your name: Build up a blog or participate in a forum, just put your name in some internet activity related to your industry.

6- Google translator: Use it! Don't be afraid of classifieds in a different language.

7.- Your 99Links: Make sure you keep a record of the links you checks.


*Good Luck!*